An Efficient Production of Bee Colonies

ISBN: 978-1-912271-74-0

Text and graphics: Flemming B. Thorsen

Translation: Bodil J. Thorsen, Roger J. Lacey and Flemming B. Thorsen

Cover photo: Roger J. Lacey

Photos: Asger S. Jørgensen, Rolf Tulstrup Theuerkauf and Flemming B. Thorsen

Layout: Rolf Tulstrup Theuerkauf

Thanks to: My dear wife Bodil Thorsen for translating from Danish to English and my dear friend Roger J. Lacey for proof reading it.

This booklet was originally published in Tidsskrift for Biavl 5/2004

English edition published by Northern Bee Books 2020
Northern Bee Books, Scout Bottom Farm,
Mytholmroyd, Hebden Bridge, HX7 5JS (UK)
www.northernbeebooks.co.uk
Tel: 01422 882751

An Efficient Production of Bee Colonies

Flemming B. Thorsen

Preface.

An Efficient Production of Bee Colonies is written by Flemming B. Thorsen, a commercial bee-keeper. He produces many new colonies every year.

He is also a commercial queen breeder.

He concentrates on breeding pure Danish Buckfast bees. There are no pure English Buckfast bees in Denmark anymore, but he continues to use the methods of Brother Adam from Buckfast Abbey in England.

Flemming also breeds pure Carnica queens as well as a combination of Buckfast and Carnica strains.

He is a member of the Danish Buckfast Breeders Association (www.avlerringen.dk)

Late Summer

Formation of 5-frame bee colonies for production of brood next year.

The absolute latest is the period 25 July – 15 August when I make 12 double colonies, each of 5 frames i.e. 24 colonies. I must be sure to have 20 queen bees that will function through one whole season, as I produce a series of 20 new colonies every 10th day. Therefore I will have 4 colonies spare. These 4 colonies are in a standby position ready to step in if one of the 20 colonies looses a queen bee. In the meantime they produce bees to be used in the queen bee mating boxes. Later during the season when the brood producing colonies are getting 'tired' and don't produce so many new bees anymore, the 4 extra colonies can also deliver bees to new colonies.

A piece of soft plastic is put over the top of the brood box and is fastened with a strip of wood on to the dividing wall (Fig.1) to prevent the bees from going from one side to the other.

After preparation of the five frame hive, each bee colony is made of: one frame of feed, and two frames of brood and working bees. They are placed next to the entrance. 1 to 21 day old bees from two frames are shaken into the hive. Bees older than 21 days will go back to their own hive. Two frames with foundation complete the series.

A mated queen bee in a cage is added. Half a packet of 2,5 kg Apifonda is placed on each colony, and the plastic cover is replaced.

Usually I place new colonies in already existing apiaries. To avoid robbing I feed the new colonies with Apifonda for 8 days. The entrances to the new colonies are restricted to 1.5 cm. After 8 days the two brood frames have emerged, the queen bee is busy laying eggs, and the colony will be able to defend itself.

The colonies are now ready to be stimulated with 50% liquid feed. The strip of wood over the piece of plastic is loosened and the plastic is removed. A double feed box is placed on the

Guide to Figures:

- foundation
- brood
- feed
- pollen and feed
- feed container or dividing wall

Figure 1.
Start: 2x5 frames colonies with young queen bees.

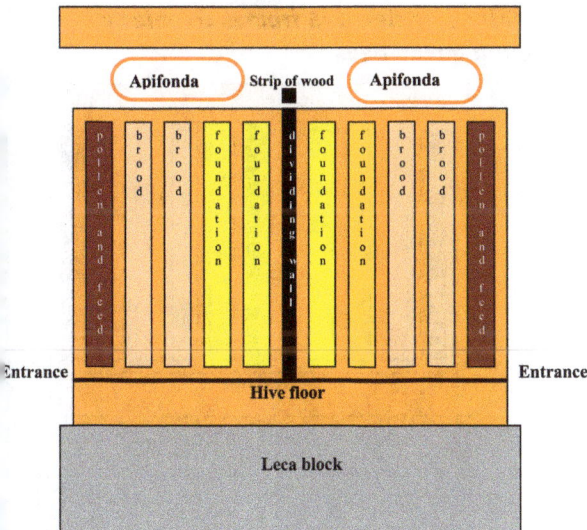

"2 colony" beehive, and the plastic is put on the feed box to keep the smell of feed from getting out again to avoid robbery. Each colony is fed with about 2 litres of 50% each (4 litres each time). The double feed box could be constructed using two feed buckets, thus preventing the two broods from amalgamating.

They are fed every 5-7 days until the 1st of October. If the bees have trouble with taking down the feed you can skip feeding once, and then go on as usual. It is important that all 5 frames are filled with feed, when you finish feeding on about October 1st. On the other hand you should not finish feeding too early as it is important that the queen bee can lay eggs for as long as she wants to. (Fig. 2 & 3 Winter colonies on 2 x 5 frames with young queen bees)

Because of the need of ventilation, the feed box ought to be removed when the colony has gathered in a cluster, and the soft plastic replaced with a cover board.

Last winter I left my feed boxes on, only the soft plastic was changed with a crown board for ventilation. The colonies stayed fine and dry.

Figure 2.
Winter Colonies: 2x5 frames colonies with young queen bees.

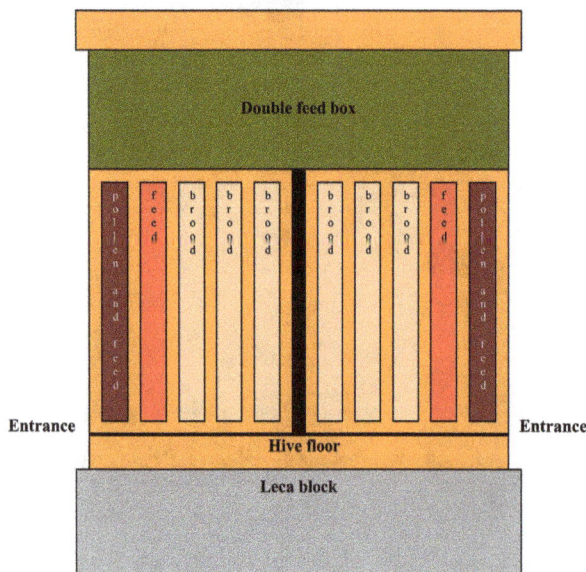

Figure 3.
Winter Colonies: 2x5 frames colonies with young queen bees.

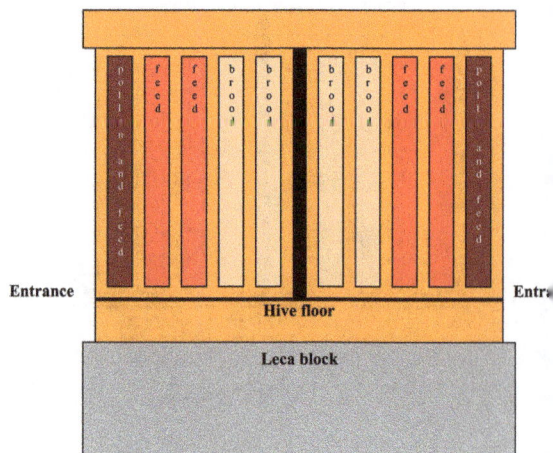

Spring

The 1st Operation (Fig 4)

Preparation of the brood producers – Schedule May 15th or 20 days before the first brood frames are to be used for new colonies. For the first operation you will need an empty "box" (hive without bottom), 8 frames of foundation, 2 feed containers, and a queen excluder.

Procedure: Take the lid off, put it on the ground, and place the empty box on it. From each side of the hive take up frames number 2 and 4 and put them in the middle of the empty box. Put 4 frames with foundation into the gaps in the hive. Put on the excluder, place this new box on top, and then put 2 frames of foundation and a feed container at both ends of the new box. In each feed container you put 1 cup of "clay" peas (to keep the bees from drowning) and 2 litres of 50% liquid feed. Five days later, again 2 litres 50% liquid feed. The 4 food and feed frames that were moved from the bottom hive to the top one are usually a mix of feed, pollen, and brood of different age.

The 4 foundation frames that were given to the queen bees, 2 each, will be drawn out and filled with eggs and larvae within 10 days.

Figure 4. 1st Operation.

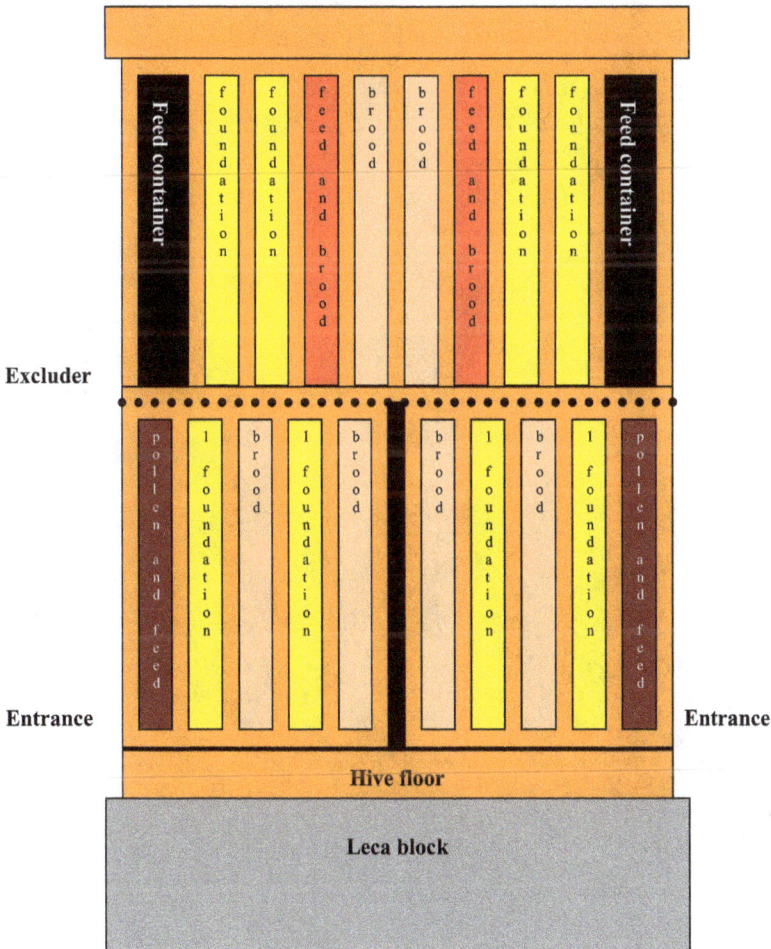

The 2nd Operation (Fig 5)

May 25th or 10 days after the first operation you do the 2nd one.

You take off the roof and put the top box on it. The 2 x 2 foundation frames next to the feed containers are taken out and placed beside the hive. The now fully sealed brood and feed frames are pushed 2 and 2 to each end of the box next to the feed containers. The excluder is removed, and again frames number 2 and 4 from each colony are taken out (the same ones that were put in during the 1st operation) and placed in the middle of the top box, (Fig. 5). **NB:** Make it a habit to place the frame with the queen bee on at the corner next to the entrance as soon as you find her, so you avoid worrying about having placed her over the excluder!

The four foundation frames from the top box are placed in the two colonies as no. 2 and 4. If the four frames should happen to be too "drawn out" and contain feed, it may be necessary to use new foundation frames instead, like you did in the 1st operation. If that's the case you take a new box filled with new foundation frames. Take from this box 4 foundation frames and put them as no. 2 and 4 in the two colonies in the bottom hive, and the 4 "built out" frames are put in the middle of the new 3rd box (Fig. 6). You feed with 2 litres 50% feed in each feed container. This is repeated 5 days later.

Figure 6.

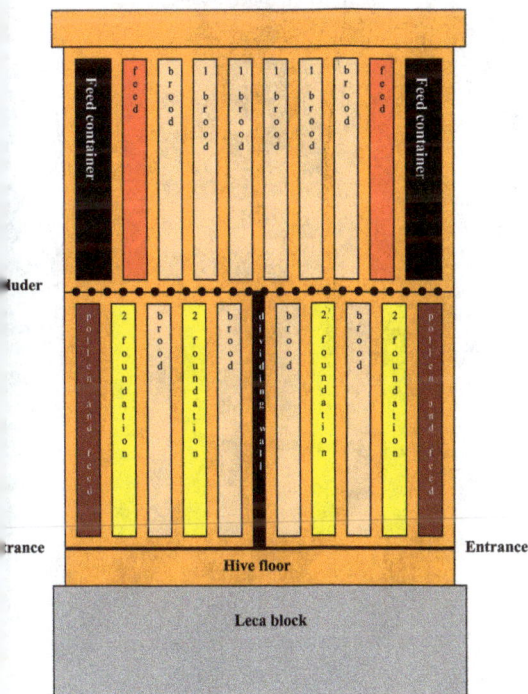

Figure 5. The 2nd Operation.

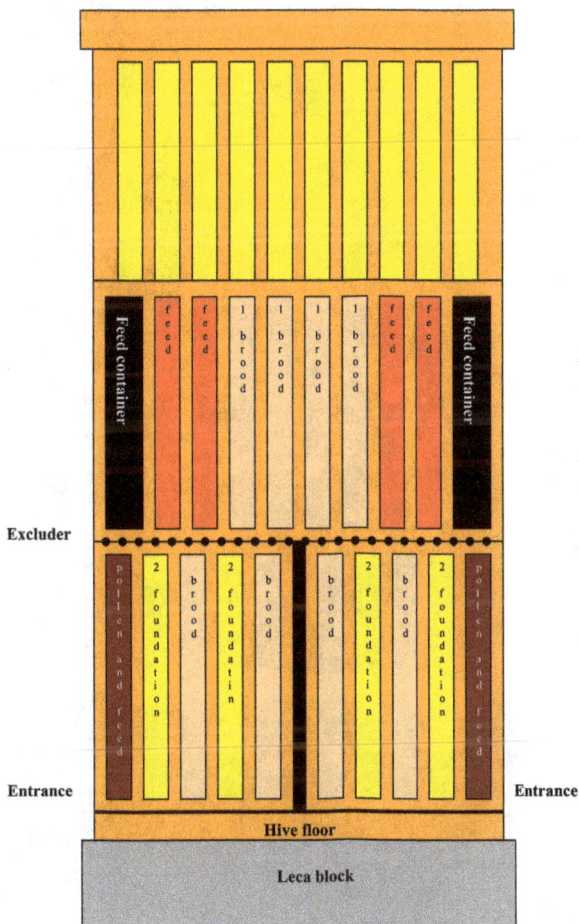

Summer

June 4th, or 20 days after the 1st operation the brood in the box over the excluder is ready to be used for making new colonies. For each brood producing queen bee you prepare a new hive with a closed entrance, 9 foundation frames, 1 feed container, a piece of plastic instead of a crown board, and a hive strap. I use the piece of plastic instead of the crown board to keep the smell of feed in the hive, so I can avoid robbery. Remember to have your new apiary ready for your new colonies.

The ready-made beehives are placed in a row behind the row of beehives with brood producing ones. The strap is loosened, lid and plastic are put aside, 5 foundation frames are taken out, 2 of them are placed beside the new hive, 3 of

them are placed next to the colony you will take the brood from. I make 20 new colonies at a time, but, if you are a beginner, I would advise you to make 2 new colonies at a time.

Making the New Colonies

From the box over the excluder you take 2 sealed brood frames and 1 feed frame and put them into the new hive next to the entrance. With a little practice you can take the 3 frames in one grip, (Fig. 7, the green arrows).The frame next to the feed container is lifted up, and you shake the bees off into the new colony and put it back again. As the two brood frames are ready to emerge (it may just have started) you don't need many more extra

Figure 7.

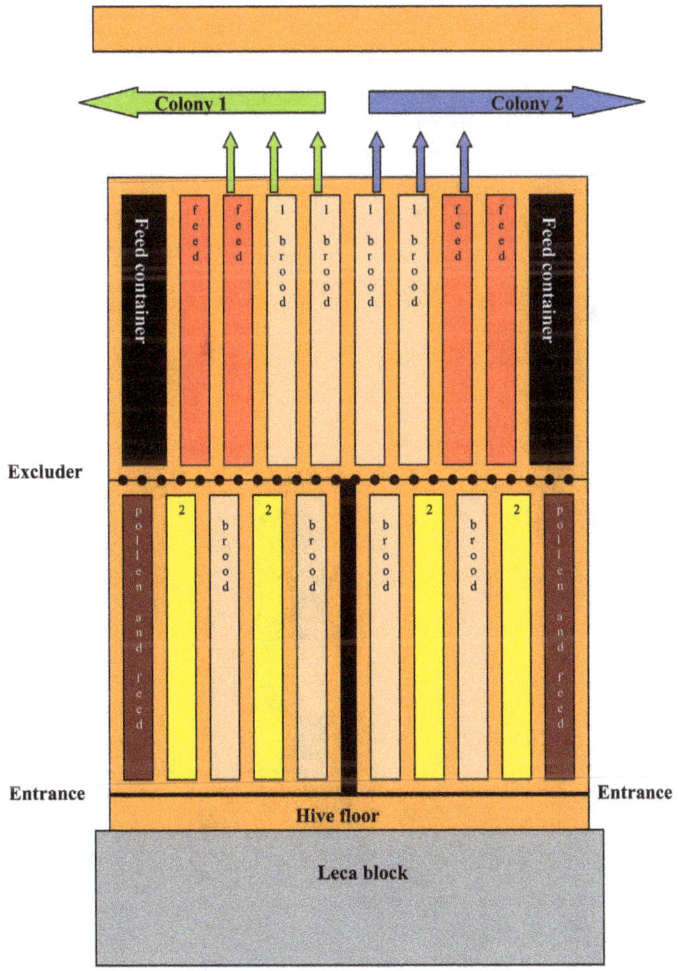

bees than the ones that were on the frames when you lifted them over. But if the third box is on the brood producer it would be a good thing to shake bees from 2 frames into the new hive. Half a litre of bees is enough. In the new hive behind the feed frame and two brood frames you now put the 2 foundation frames that were lifted out at the beginning of this operation, then the feed container, and behind it you have 4 foundation frames to be used later. Then the beehive is ready to be closed with soft plastic cover, roof and a hive strap and to be moved to the new apiary (Fig.8).

The second new bee colony is made in the same way from the other end of the brood producer (Fig. 7 blue). Remember to shake bees from the frame next to the feed container into the new colony, and if it is possible also from the third box. It is important also to leave enough bees for the second colony.

Because I make so many bee colonies at a time I choose to concentrate about getting this job done and moving them to a new apiary. Therefore I just put 6 foundation frames into the brood producer over the excluder, and finish with the sheet of plastic and roof. I'll return later to finish my 3rd operation.

In the New Apiary

The new colonies must have queen bees. For the first colonies (June 4th) you probably don't have mated queen bees at your disposal, so you have to give them each a virgin queen bee in a cage. Instead I have often bought ready-to-hatch queen cells to put into the first series of colonies. It gives 100% acceptance of the newly hatched queen bee. When I can get hold of mated queen bees (about June 15th) I use them for the new colonies.

Each new beehive receives half a packet of Apifonda, the piece of plastic is put on top, then the roof and clips (or a hive strap). The entrance is opened only 1,5 cm to avoid robbery.

After about 8 days the feed has been eaten. The 2 foundation frames have been built out. Now you can start power feeding your new colonies. 1 or 2 foundation frames are placed in front of the feed container depending on the bees' need to build. Into the feed container in which you have a cup of "clay peas" as a "float layer" you pour about 2 litres of 50% feed to make the bees work hard. This is repeated every 5th – 7th day; but if the bees collect nectar you have to cut down on feed. It is important that the queen bee is not prevented from laying eggs by too much stores, she ought to end

Figure 8.

Colony 1

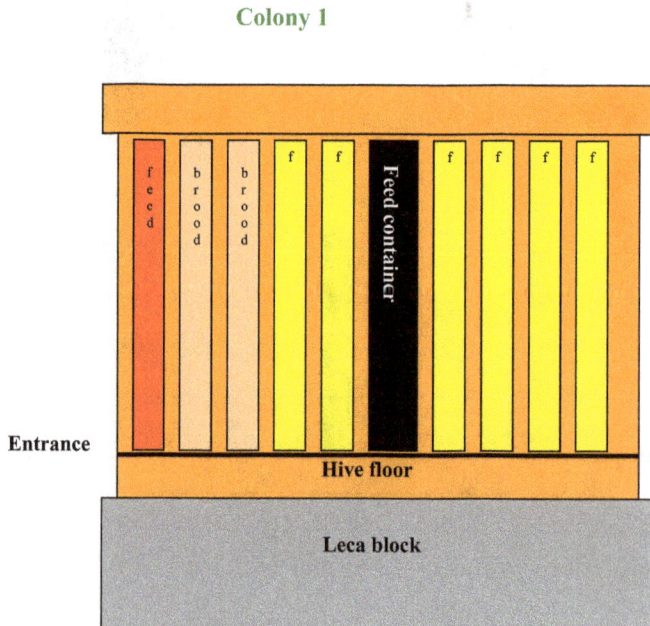

Entrance

Hive floor

Leca block

up by having 5 – 7 frames of brood. But to produce so much brood it is necessary that there are open feed cells at her disposal around the brood all the time – too much – too little ruins everything!

If you have used a virgin queen bee or a queen cell, and the weather has been right for mating you can check for eggs at the second feeding. If you loose a queen bee, which seldom happens, then use a queen cell the second time - it is far more successful. Perhaps it would be a good idea to give the colony another sealed brood frame so the colony doesn't become too small. The feeding goes on until the colony is ready for winter. If the colonies you made first have grown too big/strong, and you have enough mated queen bees at your disposal you can divide them at the end of July.

Back to the 3rd Operation in the Brood Box
From each queen bee you lift out frames no. 2 and 4. If the feeding has been done correctly the frames will be full of eggs and larvae right out to the corners. The brood frames are placed in the middle of the box. Then you lift out a pollen frame (perhaps a mixed pollen and brood frame) from each queen bee. The 2 pollen frames are placed on both sides of the brood frames. Now each queen bee receives 3 foundation frames. Check if there is open feed in one of the two remaining old frames, if not one of the three foundation frames is exchanged with half a filled feed frame from the 3rd box. This is to ensure the queen has sufficient cells for egg laying.

This frame is placed as No. 1 or 5, i.e. at the front or back of the box. The colony is fed with 2 litres of 50% power feed in each feed container. Remember to repeat the feeding every 5th day.

The 4th Operation
On June 14th the procedure of producing new bee colonies can start again with lifting out brood, feeding etc. The advantage of this system is that you don't use half or full frames of good honey as feed for your new colonies. Of course there is honey in the feed frames; but most of it will be liquid feed.

Remember: It is important that the bees have pollen at their disposal in the box over the excluder to feed the new larvae.

Remember: It is important that each queen bee under the excluder has an open feed frame at its disposal, as a rule about half a frame of open feed. Usually it is easy to find in the 3rd box.

If your brood frames are not filled with eggs and larvae right out to the corners it is seldom because of lack of pollen. It is far more important that there is open feed in the vicinity of the queen.

If your brood frames have not been built right out to the corner, your bees will probably be short of food. Conversely you feed too much, or the bees collect too much nectar themselves, they will put feed into the top corners in the new foundation frames instead of eggs and brood.

I have to mention that it can be difficult to get the frames filled at the end of the season. Finally the queen bee may have run out of eggs, or she is just not good enough at it any longer. That is why I always keep queen bees in reserve. Remember to order queen bees so you always have them ready for new colonies. I solve this problem by having a row of mating boxes with mated queens. I supply continuously during the whole summer, and in August I end up by having mating boxes with new mated queen bees to replace old ones in old colonies.

If, in the meantime, you should have forgotten: Remember to feed your new colonies every 5 to 7 days.

As brood producers I always use recently made colonies with new queen bees on clean equipment (notice July 24th). I do this to make sure that all my equipment is regularly changed and cleaned.

Wintering of Used Brood Producers

What does one do with twelve double colonies from which the queen bees have delivered 6 x 2 frames of brood for new colonies? This will be a total of about 72.000 bees per queen bee, if the frames have been built out properly.

I check up on the queen bees once more, especially those in which the egg laying capacity is reducing. If both queen bees are not doing well they are replaced, and the hive is reduced to 2 x 5 frames. Two new queen bees are put in, and a double feed box is placed on top. If only one queen bee is poor it is replaced. The number of bees will decide whether the colony is to have 10 frames with a dividing wall in the second box too, or perhaps 9 frames with feed containers at each side.

If the number of bees are too few for 10 frames you go down to 2 x 5 frames with a double food box. The used brood producers are fed slowly to make them stronger. Feeding should be at least every 7th – 10th day with 2 litres of 50% power feed per queen bee to stimulate the production of brood. The feeding continues until the 1st of October.

What will I use these colonies for next year? Read the next chapter.

Figure 9. Wintering of "used" brood colonies, ready to be used for optimizing nectar collection in spring with two brood boxes.
Winter boxes: Colonies with 2 x 10 frames (5 frames above each other) with one old queen bee and one young queen bee.

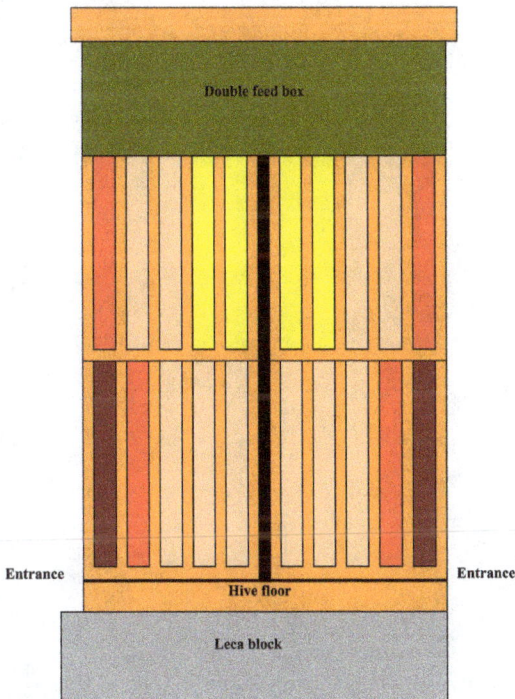

Figure 10. Winter boxes: Colonies of 2 x 5 frames with a young queen bee at the back and an old one in the other. The old queen bee is killed when the first rape flower comes out. The dividing wall is removed.

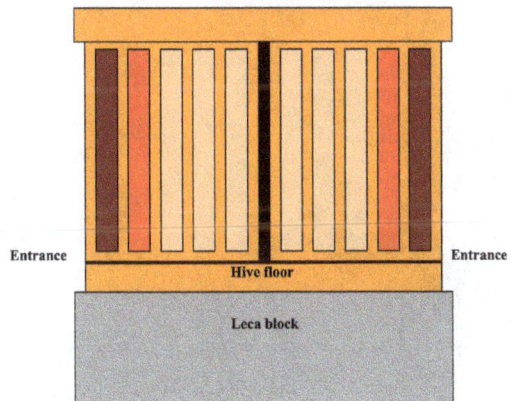

Optimizing the Collection of Nectar in Spring
Specially on Winter Rape

A few years ago we had a series of queen bees which were a little late in maturing. They were not able to get ready for the winter rape. As an experiment we placed two colonies on top of each other with an excluder and a sheet of newspaper in between. A third box with foundation frames was put on top to make sure the queen bee in the second box had sufficient space for egg laying (Fig. 11). If it was necessary another excluder and box No. 4 for honey would be put on. This system worked very well, and we didn't loose any queen bees. Later a few of the colonies were sold to be used for collecting heather nectar, for which they were very suitable because of their late brood curve.

All beekeepers know that a small colony doesn't produce any honey. Two smaller ones put together gave a reasonable yield. But - there were still two colonies to look after, also in the spring

season. This brings us back to Fig. 9. If I have brood producers from the year before with an old and a new queen bee in each side of the hive I kill the old one and take out the dividing walls when the first rape flower comes out

Usually it is easy to unite colonies at this time of the year. Now there is only one queen bee to lay eggs, and after ten days, when the rape is in full bloom, all the brood of the old queen will be sealed. By now there is just one brood area to look after and two brood areas from which young bees emerge, plus the nectar collecting bees of the two queens. This will produce honey if only we get the good weather at the right time.

Figure 10: 2 x 5 frames function more or less in the same way. If they develop well in spring it may be necessary to put on an excluder and a box to be able to move up brood frames like it was done in the brood producers. The old queen bee is kil-

Figure 11.

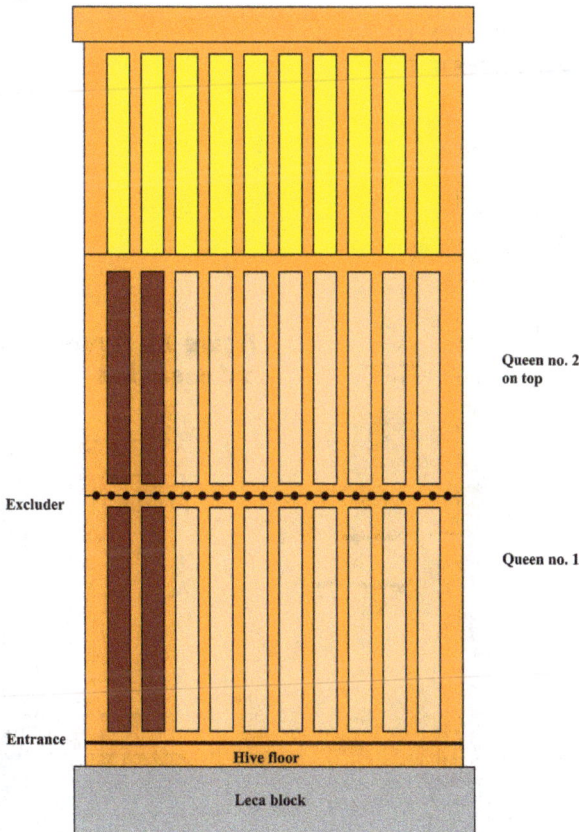

Queen no. 2 on top

Excluder

Queen no. 1

Entrance

Hive floor

Leca block

led when the first rape flower comes out, and the colonies are united.

This year we hope to make this method even more intensive as we plan to put two big colonies together using this model. The old queen bee is removed, and her brood is placed on top of the brood of a young one with a sheet of newspaper between (fig. 12). If necessary, the colony can be strengthened by transferring brood frames from other colonies.

Generally our bee colonies don't swarm very often; but it will be interesting to see what happens when they've finished collecting rape nectar in spring, and there's nothing to find anywhere else. Will we have to divide our colonies in two to make them stay at home?

Until now I've written that I kill the old queen bee. But if she's doing ok, you may choose to put her into a new hive with one feed frame, one brood frame with the bees on, one frame with foundation, and a feed container. Behind the container you fill out the gap with foundation frames to be used later.

Power feeding every 5th – 7th day, at the beginning with just 1 litre. Let the colony sit a little tight until it has got going. To avoid robbery narrow the entrance to 1,5 cm. In the late summer you will have a good colony ready to receive a new mated queen bee. If it is a strong colony, you might divide it into two 5 frame ones with two new queen bees.

Figure 12. Putting together two colonies, when the first rape flower comes out.

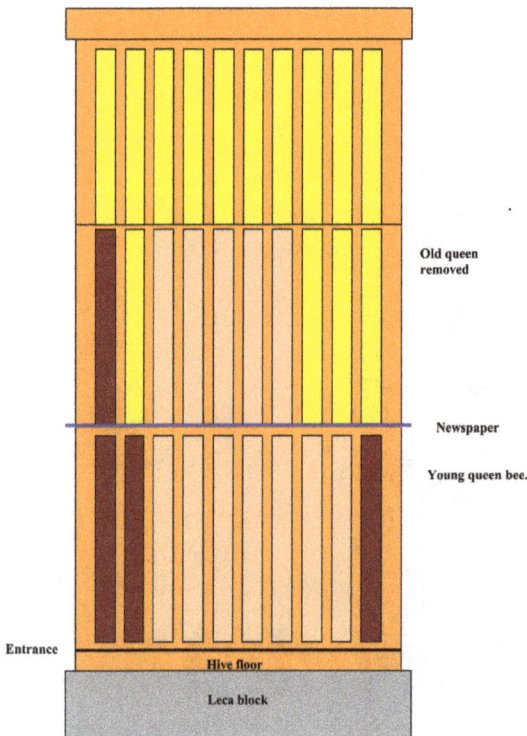

Old queen removed

Newspaper

Young queen bee.

Entrance

Hive floor

Leca block

Figure 13. A recently made colony with an old queen bee.

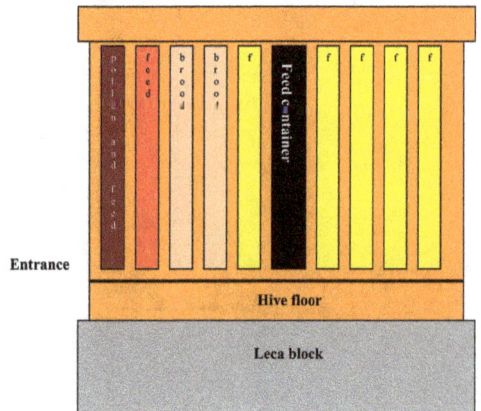

Entrance

Hive floor

Leca block

A Calendar

May 15, The First Operation
- mixed feed and brood frames are moved up over the excluder, feeding
- May 20, feeding

May 25, The Second Operation
- new brood frames up, foundation frames down, feeding
- May 30, feeding

June 4, The Third Operation
- the first new colonies are made
- brood frames up, check up on pollen over the excluder
- foundation frames down, check up on open feed under the excluder, feeding
- June 9, feeding

June 14, The Fourth Operation
- new colonies are made
- brood frames up, check up on pollen
- foundation frames down, check up on feed, feeding
- June 19, feeding

June 24, The Fifth Operation
- new colonies are made
- brood frames up, check up on pollen
- foundation frames down, check up on feed, feeding
- June 29, feeding

July 4, The Sixth Operation
- new colonies are made
- brood frames up, check up on pollen
- foundation frames down, check up on feed, feeding
- July 9, feeding

July 14, The Seventh Operation
- As above

July 24, The Eighth Operation
- formation of new five frame colonies, which are going to be brood producers next year
- look at figure 1 and the description of the procedure at the beginning of the article.